Soil pollution

Dr. Hemant Pathak

ISBN: 1492762369
ISBN-13: 978- 1492762362

DEDICATION

Dedicated to Shri Sainath Maharaj the all omnipotent of world the most merciful.

CONTENTS

Foreword

Soil pollution; provides a unique insight into the problems our planet faces in terms of clean environment, and what to do about it. This is the only books Written for academics, researchers and practitioners working in soil pollution and management field, expressed comprehensive and interdisciplinary focus on the ecological issues associated with Soil pollution to provide a complete picture of current environmental problem from cause to effect to solution

This book made of 10 years consistently research on environmental issues, makes it ideal source for students, teachers, industrialist, environmental experts and economists.

This book provides an essential guide to researchers, it offers: various causes of pollution; on the challenges and experiences in present scenario.

Simply explained, Soil pollution is an important book bringing together diverse viewpoints from academia and environmental agencies and regulators, for all who wish to make a difference in how to plan and manage our Environmental resources.

Dr. Hemant Pathak

M.Sc. (Gold medalist), Ph. D.

Assistant Professor of Engineering Chemistry

Indira Gandhi Govt. Engineering College,

Sagar, MP, India

Glossary

Abatement   The reduction or elimination of pollution.

Acid rain   The precipitation of dilute solutions of strong mineral acids, formed by the mixing in the atmosphere of various industrial pollutants

Act   A law

Aerated soil   An aerated soil is a soil with a good movement of air through the soil structure. The opposite is a wet waterlogged soil.

Aggregates   Soil aggregates are soil 'lumps' of a range of sizes.

Aerosol   Particles of solid or liquid matter than can remain suspended in air from a few minutes to many months depending on the particle size and weight.

Anaerobic soils   Anaerobic soils have very little oxygen present.

Ash   Incombustible residue left over after incineration or other thermal processes.

Bearing capacity   This is effectively the weight a soil can withstand before severe damage occurs to the structure of the soil. Bearing capacity varies throughout the year, for instance a very heavy tractor that causes no damage on dry soils may cause a lot of damage to the soil structure of wetted soils.

Biodiversity   A large number and wide range of species of animals, plants, fungi, and microorganisms. Ecologically, wide biodiversity is conducive to the development of all species.

Clay   That mineral fraction of the soil with particles smaller than 0.002 mm in diameter.

Combustion   Burning. Many important pollutants, such as sulfur dioxide, nitrogen oxides, and particulates (PM-10) are combustion products, often products of the burning of fuels such as coal, oil, gas, and wood.

Contamination   The act of polluting or making impure; any indication of chemical, sediment, or biological impurities.

Denitrification The process by which nitrate-nitrogen is converted to nitrogen gas by soil microorganisms when soil oxygen is low or absent.

Dust Solid particulate matter that can become airborne.

Erosion Erosion is the wearing away of land or soil through one or more processes. Causes of erosion include the actions of water (rills, inter-rill, gully, snowmelt and river and lake bank erosion), wind, translocation and geological.

Effluent Municipal sewage or industrial liquid waste (untreated, partially treated, or completely treated) that flows out of a treatment plant, septic system, pipe, etc.

Evaporation This is the rate of water loss from liquid to vapour (gaseous) state from an open water,wet soil or plant surface, usually expressed in mm day-1.

Geology Geology is a scientific field concerning the study of rock.

Horizon One of the layers that form in the soil profile as a result of soil-forming processes. A horizon can appear as a marked visible layer, more usually horizons boundaries are more subtle.

Humus Organic matter (humus) forms from the decay of leaves, plants and other life.

Infiltration Infiltration is the movement of water from the surface down into the soil before moving down to the aquifers, or out to rivers. A portion of soil water may also be lost via the process of evapotranspiration.

Hydrocarbons Compounds containing various combinations of hydrogen and carbon atoms. They may be emitted into the air by natural sources (e.g., trees) and as a result of fossil and vegetative fuel combustion, fuel volatilization, and solvent use. Hydrocarbons are a major contributor to smog.

Leaching Leaching is the process where soluble materials (including nutrients and salts) in the soil are washed down the soil profile by water.

Pan	A pan is a well-defined layer forming in the soil.

Parent material Soil parent material refers to the rocks which were weathered to form the soil in the first place. Usually the parent material is below the soil, but it can be distant if glaciers translocated the soils during the ice ages. Parent material is the focus of the study of geology.

Pedology The science of studying soils and their interaction with the wider environment.

pH pH is a measure of acidity; It is measured from 1 (acid) through 7 (neutral) to 14 (alkaline) expressed on a logarithmic scale. Most soil is about pH 3 to 8.

Pores A soil pore is the hole in-between particles of soil that can become filled with air or water.

Precipitation Precipitation means water reaching the ground from both rainfall, snow and hail.

Profile The soil profile is a column of soil, essentially three-dimensional and large enough to be used to characterise the soil condition at a particular place

Pollution control The addition of processes, practices, materials, products or energy to waste streams to reduce the risk posed by pollutants and waste before their release to the environment.

Pollution Prevention The use of processes, practices, materials, products, substances or energy that avoid or minimize the creation of pollutants and waste, and reduce the overall risk to human health or the environment

Public health the health or physical well-being of a whole community.

Runoff Runoff occurs as water falling as precipitation does not soak deep into the soil, but passes across the surface and through the near-surface towards the rivers.

Soil	Soil is made from the breaking down of rocks and organic matter by physical, chemical and biological processes.
Soils Acidity	Most soils are of a pH from about 5.5 to 8, this is a large range but some soils can be pH 3, which is very acidic.
Soil minerals	The trace elements found in soil (nutrients)
Soil structure	Architecture of soil. structure is the aggregation of primary soil particles into units separated from each other by surfaces of weakness.
Reuse	The reemployment of products or materials, in their original form or in new applications, with refurbishing to original or new specifications as required.
Risk assessment	Methods used to quantify risks to human health and the environment.
Texture	The description of the balance in the soil between the constituents including sand, silt and clay as well as organic matter.
Topsoil	Topsoil is the surface layer of soil containing partly decomposed organic debris, and usually high in nutrients, containing many seeds, dark colour due to the 'organic matter' present.
Water holding capacity	Soil organic matter increases the water holding capacity. Pure running sand has a low water holding capacity.
Weathering	The process by which materials are broken down into smaller parts and ultimately their constituents by physical, chemical and biological weathering processes.

1. Introduction

Soil is one of the important and valuable resources of the nature. It consist of thin layer of organic and inorganic materials that covers the Earth's surface. Life and living on the earth would be impossible without healthy soil. The soil as a place where ecological functions are conducted that are relevant for the biosphere.

95% of human food is derived from the earth, it filter and buffer, allowing clean groundwater. Healthy and productive soil is essential to human survival source of ores and physical base for construction works.

Contamination of soil system by considerable quantity of chemicals or other substances resulting in reduction of its fertility. Entrance of materials, biological organisms or energy into the soil will cause changes in soil quality. Productive soils are necessary for agriculture to supply the world with sufficient food.

Soil pollution is caused by the Increasing urbanization, presence of man-made chemicals contamination typically arises from the rupture of underground storage links, discharge of domestic and industrial waste into the soil, disposal of untreated wastes, Unscientific mining Seepage from a landfill Solid waste, over uses of agrochemicals & agricultural application of pesticides, herbicides or fertilizer, percolation of contaminated surface water to subsurface strata, oil and fuel dumping, leaching of wastes from landfills or direct discharge of industrial wastes to the soil. Outdated technology, inadequate treatment and safety management of chemicals and waste. The most common chemicals involved are petroleum hydrocarbons, solvents, pesticides, lead and other heavy metals.

Soil pollution comprises the pollution of soils with materials, mostly chemicals, that are out of place or are present at concentrations higher than normal which may have adverse effects on humans or other organisms.

The presence of any element in a fatal concentration in the soil due to natural and of agricultural land with application of pesticides will result in heavy metals such as copper, nickel, zinc and cadmium accumulating in the topsoil. Toxic ions of aluminum, zinc, copper, lead, cadmium and other metals are often accumulated in the degraded soils of industrial regions. toxic ions inhibit root development and at higher concentrations lead to root necrosis or death.

The effects of soil pollution are very hazardous and can lead to the loss of ecosystems. When soil is polluted, it directly or indirectly affects the climate patterns.

2. Composition of soil

Soil is composed of soil living part and dead part. Organic portion, which is derived from the decayed remains of plants and animals, is concentrated in the dark uppermost topsoil. inorganic portion made up of rock fragments, formed over millions of years by physical and chemical weathering of bedrock. Soil properties vary from place to place and are influenced by many factors. The permeability of soil formations underlying a waste disposal site is of great importance with regard to land pollution. The greater the permeability, the greater the risks from land pollution.

Table.1 : Content of Various Elements in Soils

Metals	Common Range for Soils mg/kg
Al	10,000- 300,000
Fe	7,000-550,000
Mn	20-3,000
Cu	2-100
Cr	1-1000
Cd	0.01-0.70
Zn	10-300
As	1.0-50
Se	0.1-2
Ni	5-500
Ag	0.01-5
Pb	2-200
Hg	0.01-0.3

3. Sources of Soil Pollution

There are many different sources of soil pollution. Indiscriminate discharge of industrial effluents on land and into water bodies. commercial and industrial chemicals cause contamination through accidental spills or leaks. Petrol pump and mechanics garages use different fuels and lubricants on-site. These contaminants generally enter the soil inadvertently as a result of poor storage practices or spillage onto the ground. Open defecation by animals and human beings. Waste disposal sites are one of the major sources of elevated levels of metals in the soil environment. Migration of contaminants from waste disposal sites to surrounding ecosystems is a complex process and involves various geochemical activities.

Due to urbanization, large amount of construction activities are taking place which has resulted in large waste articles.

Contaminants can also be spread through the air and are deposited as dust or by precipitation. During extraction and mining activities, several land spaces are created beneath the surface. Radioactive substances from nuclear plants which are released into the soil.

Large amount of solid waste is leftover once the sewage has been treated. The leftover material is sent to landfill site which end up in polluting the environment.

fertilizers and Pesticides added to soils to get rid of insects, fungi and bacteria from their crops. The overuse of these chemicals, they result in contamination and poisoning of soil.

Plants obtain carbon, hydrogen and oxygen from air and water. Soil nutrients like nitrogen, phosphorus, potassium, calcium, magnesium, sulfur are important for plant growth and development. Farmers use fertilizers to fulfill soil needs.

Fertilizers contaminate the soil with impurities, which come from the raw materials used for their manufacture. As, Pb and Cd present in traces in rock phosphate mineral get transferred to super phosphate fertilizer. Since the metals are not degradable, their accumulation in the soil above their toxic levels due to excessive use of phosphate fertilizers, becomes an indestructible poison for crops.

With growing human population, demand for food has increased considerably. Plants on which we depend for food are under attack from insects, fungi, bacteria, viruses, rodents and other animals. To kill unwanted populations living in or on their crops, farmers use pesticides.

Since it was soluble in fat rather than water, it biomagnified up the food chain and disrupted calcium metabolism in birds, causing eggshells to be thin and fragile. As a result, large birds of prey such as the brown pelican, ospreys, falcons and eagles became endangered.

The most important pesticides are DDT, BHC, chlorinated hydrocarbons, organophosphates, aldrin, malathion, dieldrin, furodan, etc.

The remnants of such pesticides used on pests may get adsorbed by the soil particles, which then contaminate root crops grown in that soil. The consumption of such crops causes the pesticides remnants to enter human biological systems, affecting them adversely.

4. Chemicals causing soil pollution

•Metallic pollutants-textiles, dyes, soaps, detergents, drugs, cement, rubber, paper, metal industries release Pb, Fe, Cu, Zn, Hg, Cd, CN, acids, alkalies etc.

•Agro chemicals-Fertilizers, pesticides, insecticides, rodenticides, weedicides, fumigants release toxic chemicals like Pb, As, Cd, Hg, Co etc.

•Radioactive Chemicals

5. Soil and acidity

Most of the metals are soluble in acid soils than in neutral or slightly basic soils. pH affects the solubility of metals. The solubility of metals is strongly influenced by the redox potential, the presence of complexing agents such as chlorides, sulphates, carbonates and organic acids, and the properties of the solid waste phases in or on which metals can be bound.

These metals can bio-magnify in plants and animals eventually making their way to humans through the food chain. Iron, manganese and lead have low to very low mobility at $pH < 7$ and thus, would be enriched in soil. Nickel, copper and zinc have high mobility under acidic conditions and due to formation of sparingly soluble metal sulphides with very low mobility under reducing conditions, these metals in soils can either be enriched or depleted relative to parent material depending on the dominant factors that exist in the weathering environment

6. Effect of Soil pollution

Soil pollution decreases the growth rate of trees and shrubs and decline of whole forest area. Acid precipitation caused by emissions of sulphur dioxide and nitrogen oxides is the major reason for considerable acidification of soils.

Soil pollutants would bring in alteration in the soil structure, which would lead to death of many essential organisms in it. This would also affect the larger predators and compel them to move to other places, once they lose their food supply.

Increase in acidity of rainwater and wet deposition is observed as a result of a decrease in dust alkaline components through introduction of more efficient dust removal and insufficient desulphurization in flue gases of power and heating plants. Increase in salinity of the soil, which therefore makes it unfit for vegetation, thus making it useless and barren.

The soil when contaminated with toxic chemicals and pesticides lead to problem of skin cancer and human respiratory system. The toxic chemicals can reach our body through foods and vegetables that we eat as they are grown in polluted soil.

Toxic ions of aluminum, zinc, copper, lead, cadmium and other metals are often accumulated in the degraded soils of industrial regions. Toxic ions inhibit root development and at higher concentrations lead to root necrosis or death.

Aluminium is widespread in the Earth's crust, and its availability to plants increases with

Table.2- Soil pollutants and their effect on human health

Metal	Source	Effects
Arsenic	occurs naturally	Chronic poisoning leads to a loss of appetite and weight, diarrhea, alternating with constipation, gastro intestinal problems, peripheral neuritis, conjunctivitis and skin cancer
Cadmium	Metallurgy, mining, chemical industry and electroplating	leads to chronic poisoning and affects the proximal tubules of the kidney, causing formation of kidney stones
Lead	lead smelters storage battery	lead poisoning can lead to severe mental retardation or death
Mercury	industrial wastes	neurological problems and damages renal and tubules
Cyanides	wastes from heat treatment of metals, dismantling of electroplating shops, etc.	rapid death may follow due to exposure to cyanide as a result of inhibition of cellular respiration

decreasing pH of the soil .

Creation of toxic dust is another potential effect of soil pollution causes of disturbance in the balance of flora and fauna residing in the soil.

7. **Prevention of Soil pollution**

Natural resources have to be prolonged to their completely use to maintain the aim for continual economic growth and lessen environmental impacts. This involves reducing wastage in operations, utilizing waste products through recycling and recovery practices to further ensure the long-term availability and usefulness of natural resources. continues Attempts are being made to decontaminate polluted soils, of both on-site, in the soil and off-site removal of contaminated soil for treatment techniques.

This goal may be achieved by some activities such as:

I.   Implementing better housekeeping practices to minimize leaks and fugitive releases from manufacturing processes. Make people aware about the concept of Reduce, Recycle and Reuse.

II.   Redesigning products to cause less waste or pollution during manufacture, use, or disposal altering production processes to minimize the use of toxic chemicals

III.   Reduce the use of pesticides and fertilizers in agricultural activities. Planting pest resistant crops can reduce or eliminate the need for chemical pesticides, thereby reducing the water, air, and soil pollution that results from the manufacture and use of agricultural chemicals. Encourage Organic gardening and eat organic food that will be grown without the use of pesticides. Do Organic gardening and eat organic food that will be grown without the use of pesticides.

IV.   Materials such as glass containers, plastic bags, paper, cloth etc. can be reused at domestic levels rather than being disposed, reducing solid waste pollution. In office settings, simple steps such as making double sided copies. Printing drafts on the back sides of discarded paper can substantially reduce the consumption and disposal of paper products. they will lead to garbage and end up in landfill site. recovery of one tonne of paper can save 17 trees.

V.   To minimizing the use of toxic household chemicals such as drain cleaners and herbicides will reduce the amount of hazardous chemicals that eventually end up in the environment.

VI.   Keeping the environment clean and managing the wastes with the Guide lines of respective Government. The Exhausts from the Automobiles and workshop machinery should be controlled. Repair and replacement of leaking and malfunctioning equipment.

VII.   plants are effective in cleaning up contaminated soil. Phytoremediation is a general term for using plants to remove, degrade, or contain soil pollutants such as heavy metals, pesticides, solvents, crude oil, polyaromatic hydrocarbons, and landfill leacheates.

VIII.   Materials such as paper, some kinds of plastics and glass can and are being recycled. This decreases the volume of refuse and helps in the conservation of natural resources. The ISO standards must be followed strictly for Industrial usage. Buy biodegradable products.

IX.   To used Eco-friendly means like bicycle, bike etc. Must used public transportation means like bus for routine jobs. Administration must promoted car pool to office and back.

X.   Reduce the use of aerosols in the household. Promote the afforesting. Switch-off all the lights and fans when not required. Promoted to sharing of room with others when the air conditioner, cooler or fan is on.

XI.   To use biological remedy (Bioremediation) to abate or clean up contamination. This makes it different from remedies where contaminated soil or water is removed for chemical treatment or decontamination, incineration, or burial in a landfill. Microbes are often used to remedy environmental problems found in soil, water, and sediments. Plants have also been used to assist bioremediation processes.

XII.   Applying bio-fertilizers and manures can reduce chemical fertilizer and pesticide use. Biological methods of pest control can also reduce the use of pesticides and thereby minimize soil pollution. Control of land loss and soil erosion can be attempted through restoring forest and grass cover to check wastelands, soil erosion and floods. Crop rotation or mixed cropping can improve the fertility of the land. Control of land loss and soil erosion can be attempted through restoring forest and grass cover.

8.   Soil pollution Management

The most common decontamination method for polluted soils is to remove the soil and deposit it in landfills or to incinerate it. Mix polluted soil with cement into package. Excavate contaminated soil, and then set-to the designated landfill. Compress contaminated soil and package with the container.

Soil and crop management methods can help prevent uptake of pollutants by plants, leaving them in the soil. The soil becomes the sink, breaking the soil-plant animal or human cycle through which the toxin exerts its toxic effects.

Excavate contaminated soil and replaced it with uncontaminated soil. Deeply plow into the soil underlying contaminated soil. Cover new soil on contaminated soil. By Heating contaminated soil, thermal decomposition of pollutants.

The following management practices will not remove the heavy metal contaminants, but will help to immobilize them in the soil and reduce the potential for adverse effects from the metals. By putting chemical substances into the pollutant soil, changes soil pH or occurs binding reaction with contaminated soil to produce a new non-toxic or easy to deal with new substances.

Increasing the soil pH to 6.5 or higher. Cationic metals are more soluble at lower pH levels, so increasing the pH makes them less available to plants and therefore less likely to be incorporated in their tissues and ingested by humans. Raising pH has the opposite effect on anionic elements. By the monitoring and corresponding comparison of soil quality standards, they determine the status of the pollution of the soil, for the follow-up to take the appropriate measures to provide the basis for soil remediation.

9. Bioremediation by micro-organisms

Bioremediation is the use of metabolic activity of biological life, reduce the soil concentration of toxic and hazardous compounds, so that contaminated soil back to health process. Biodegradation is facilitated by aerobic soil conditions and soil pH in the neutral range (between pH 5.5 to 8.0), with an optimum reading occurring at approximately pH 7, and a temperature in the range of 20 to 30°C.

These physical parameters can be influenced, thereby promoting the microorganisms' ability to degrade chemical contaminants. Of all the decontamination methods bioremediation appears to be the least damaging and most environmentally acceptable technique.

Organic matter, especially for the repair of oil polluted areas. Added soil nitrogen, phosphorus and other nutrients, increase oxygen and water, so that the soil microbial maintain optimum vitality, which break down crude oil.

Compost is also commonly used bioremediation that can be greatly reduced concentration of organic pollutants in soil.

10. Soil Pollution: An Indian Scenario

India's higher economic growth is increased consumption of the natural resources and increased waste generation that contributes to ecological degradation, which is estimated at around 5% of India's Gross Domestic Product (GDP).

Some of the key areas of waste generation are liquid waste, Industrial waste including hazardous wastes, municipal wastes and e waste.

Accumulation of solid waste; this is a major problem in developed countries like India where the garbage and refuse products are not degraded. India loses 20 tons of topsoil per hectare in a year due to floods, rainfall and deforestation.

20 % to 50 % of lands under irrigation can go out of cultivation at this rate because of water logging and salinity. Some important facts concerned to India is as follows:

•>125 major contaminated sites across the country

•175 million hectare (out of 329 million ha) are considered degraded

•> 40 % of chemical fertilizers leached into soil

•> 65 per cent of India's villages are exposed to residual pesticides risk

•Heavy metals beyond permissible limits affecting GW of 40 districts from 13 states

Table.3- HW Contaminated Dump Sites – Priority List

S.No	State	Name of the Area	No. of Site	Public/Private Sector	Nature of Contaminant
1	Gujarat	Vadodara	1	Private Sector Unit	Chromium
2	Kerala	Eloor-Edayar, Cochin	4	Public/Private Sector Units	Heavy metal and POPs
3	Madhya Pradesh	Ratlam	4	Private Sector Units	Gypsum, iron salts and Naphthalene
4	Orissa	Ganjam	3	Private Sector Unit	Mercury
5	Orissa	Talcher	1	Public Sector Unit	Chromium
6	Orissa	Sundergarh	4	Private Sector Unit	Chromium
7	Rajasthan	Bichhadi	1	Public/Private Sector Units	Inorganic Salts, Organics
8	Tamil Nadu	Ranipet	1	Public Sector Unit	Chromium
9	Uttar Pradesh	Rakhimandi, Kanpur	1	Unknown (Orphan)	Chromium
10	Uttar Pradesh	Rania, Kanpur Dehat	1	Unknown (Orphan)	Chromium
11	Uttar Pradesh	Lucknow	1	Private Sector Unit	HCH (hexa chloro cyclo hexane)
12	West Bengal	Nibra Village, Howrah	1	Unknown (Orphan)	Chromium

Ref.-MOEF,India

11. References

1. Abrahams PW (2002). Soils: Their implications to human health. Sci. of the Total Environ. 291: 1-32.

2. Adriano DC (1986). Trace elements in the terrestrial environment, Springer- Verlag, New York.

3. Alloway BJ (1995a). Heavy Metals in Soils, 2nd edn. pp. 368. Blackie Academic and Professional, London. ISBN 0-7514-0198-6.

4. Alloway BJ (1995b). Cadmium. Heavy metals in soils, 2nd edn. pp.122-151. (Ed. Alloway BJ), Blackie Academic and Professional. Glasgow, U.K.

5. Bozkurt S, Moreno L, Neretnieks I (2002). Long-term processes in waste deposits. Sci. Total Environ. 250: 101-121.

6. CCME (1999). Canadian Council of Ministers of the Environment updated 2001. Canadian Soil Quality Guidelines for the Protection of Environmental and Human Health. Canadian Environmental Quality Guidelines, No.1299. CCME, Winnipeg. ISBN 1-896997-34-1.

7. Craul PJ (1985). A description of urban soils and their desired characteristics. J. Arboricult, 2: 330-339.

8. Gnaneshwar P, Sitaramayya S (1998). Petrochemistry and origin of archean granitic rocks of Hyderabad city. Indian J. Geol. 70(3): 249-264.

9. Govil PK (1985). X-ray fluorescence analysis of major, minor and selected trace elements in new IWG reference rock samples. J. Geol. Society of India 26: 38-42.

10. Govil PK (1993). Reference rock standards: preparation and evaluation. NGRI-D, Applications of analytical instruments. pp.140-145. Allied Publishers.

11. Govil PK, Narayana BL (1999). New reference material of dunite rock: NGRI-UMR: Preparation and evaluation. Geostandards Newsletter. J. Geostand. Geoanal. 23: 77-85.

12. Govindaraju K (1994). Compilation of working values and sample description for 272 Geostandards. pp.158. Geostandards Newsletter.

13. Janardhan RY (1965). The origin of Hyderabad granites-A new interpretation. J. Indian Geosci. Assoc. 5: 111-118.

14. Jung CH, Matsuto T, Tanaka N, (2006). Flow analysis of metals in a municipal solid waste management system. J. Waste Manage. 26: 1337-1348.

15. Kanungo DN, Rama Rao P, Murthy DSN, Ramana Rao AV (1975). Structural features of granites around Hyderabad, Andhra Pradesh. Geophysical Res. Bul., 13(3-4): 337-357.

16. Kiekens L (1995). Zinc. Heavy metals in soils, 2nd edn. pp. 284-305. (Ed. Alloway, BJ), Blackie Academic and Professional. Glasgow, U.K.

17. Lindsay WL (1979). Chemical equilibria in soils. John Wiley and Sons, New York.

18. Mattigod SV, Page AL, (1983). Assessment of metal pollution in soils. Applied Environmental Geochemistry (ed. Thornton, I) Chap.12, pp. 355-394. Academic Press.

19. Mc-Bride MB (1994). Trace and toxic elements in soils. Environmental Chemistry of Soils, New York, Oxford University Press.

20. Mc Grath SP (1995). Chromium and nickel, Heavy metals in soils, 2nd edn. pp.152-178. (Ed. Alloway, BJ), Blackie Academic and Professional. Glasgow, U.K.

21. Murthy NN (2008). A case study of soil contamination at Katedan Industrial Area, Hyderabad. NGRI Proc.; Contaminated groundwater monitoring and soil assessment. pp. 1-12. Hyderabad, India.

22. Nicholson FA, Smith SR, Alloway BJ (2003). An inventory of heavy metals inputs to agricultural soils in England and Wales. Sci. Total Environ. 311: 205-219.

23. Palumbo B, Angelone M, Bellanca A (2000). Influence of inheritance and pedogenesis on heavy metal distribution in soils of Sicily, Italy. Geoderma 95 (3-4): 247-266.

24. Pandey OP, Agrawal PK, Chetty TRK (2002). Unusual lithospheric structure beneath the Hyderabad granitic region, eastern Dharwar craton, south India. Phys. Earth and Planetary Inter. 130: 59-69.

25. Rao TG, Govil PK, (1995). Merits of using Barium as a heavy absorber in major element analysis of rock samples by XRF: New data on ASK-1 and ASK-2 reference samples, Analyst. 120: 1279-1282.

26. Salonen V, Korkka-Niemi K (2007). Influence of parent sediments on the concentration of heavy metals in urban and suburban soils in Turku, Finland. Appl. Geochem. 22: 906-918.

27. Schmitt HS, Sticker H (1991). Heavy metal compounds in the soil. Merian, E. (Ed.), Metals and their compounds in the Environment. pp. 311-326. Weinheim VCH.

28. Shanker AK, Cervantes C, Loza-Tavera H, Avudainayagamdet S (2005). Chromium toxicity in plants. Environ. Int. 31: 739-753.

29. Soil survey manual (1993). Soil Conservation Service. U.S. Department of Agriculture Handbook No.18.

30. Domsch KH (1984) Effects of pesticides and heavy metals on biological processes in soil. Plant and Soil 76: 367–378

31. Doran JW & Safley M (1997) Defining and assessing soil health and sustainable productivity. In: Pankhurst CE, Doube BM & Gupta VVSR (Eds) Biological Indicators of Soil Health (pp 1–28). CAB International, Wallingford

32. Edwards CA, Subler S, Chen SK & Bogomolov DM (1996) Essential criteria for selecting bioindicator species, processes, or systems to assess the environmental impact of chemicals on soil ecosystems. In: Van Straalen NM&Krivolutsky DA (Eds) Bioindicator Systems for Soil Pollution (pp 67–84). Kluwer Academic Publishers, Dordrecht

33. Eijsackers H (1994) Ecotoxicology of soil organisms: Seeking the way through a pitch-dark labyrinth. In: Donker MH, Eijsackers H & Heimbach F (Eds) Ecotoxicology of Soil Organisms (pp 3–32). Lewis Publishers, Boca Raton

34. Eijsackers H (1998) Soil quality assessment in an international perspective: Generic and land-use based quality standards. Ambio 27: 70–77

35. Elliot ET (1997) Rationale for developing bioindicators of soil health. In: Pankhurst CE, Doube BM & Gupta VVSR (Eds) Biological Indicators of Soil Health (pp 49–78). CAB International, Wallingford

36. Heimbach F (1992) Effects of pesticides on earthworm populations: Comparison of results from laboratory and field tests.In: Greig-Smith PW, Becker H, Edwards PJ & Heimbach F (Eds) Ecotoxicology of Earthworms (pp 100–106). Intercept Ltd., Andover

37. Karr JR (1992) Ecological integrity: Protecting earth's life support systems. In: Costanza R, Norton BG & Haskell BD (Eds) Ecosystem Health. Key Goals for Environmental Management (pp223–238). Island Press, Washington, D.C.

38. Lindsay WL (1979). Chemical equilibria in soils. John Wiley and Sons, New York.

39. Abrahams PW (2002). Soils: Their implications to human health. Sci. of the Total Environ. 291: 1-32.

40. Adriano DC (1986). Trace elements in the terrestrial environment,Springer- Verlag, New York.

41. Alloway BJ (1995a). Heavy Metals in Soils, 2nd edn. pp. 368. Blackie Academic and Professional, London. ISBN 0-7514-0198-6.

42. Alloway BJ (1995b). Cadmium. Heavy metals in soils, 2nd edn. pp.122-151. (Ed. Alloway BJ), Blackie Academic and Professional. Glasgow,U.K.

43. Craul PJ (1985). A description of urban soils and their desired characteristics. J. Arboricult, 2: 330-339

44. Mc-Bride MB (1994). Trace and toxic elements in soils. Environmental Chemistry of Soils, New York, Oxford University Press.

45. Vandana Parth (2011), Assessment of heavy metal contamination in soil around hazardous waste disposal sites in Hyderabad city (India): natural and anthropogenic implications, Journal of Environmental Research and Management Vol.2(2). pp. 027-034,

ABOUT THE AUTHOR

Dr. Hemant Pathak held positions as Assistant Professor in the department of chemistry, Govt. Indira Gandhi Engineering College, Sagar, MP, India. He had extensive experience in teaching, research and administrative management.

Dr. Pathak received his Ph.D. degree in chemistry from Dr. Hari Singh Gour Central University, Sagar, India and M.Sc. Gold medalist from Jiwaji University, Gwalior. He has published 14 books and more than 50 research papers in reputed International and National journals and received several awards. He is a member of editorial boards and reviewer boards of several international journals and societies. His area of specialization includes Engineering Chemistry and Environmental Pollution management.